The Cheviot Sheep

by C.S. Plumb

with an introduction by Jackson Chambers

Self Reliance Books

Get more historic titles on animal and stock breeding, gardening and old fashioned skills by visiting us at:

http://selfreliancebooks.blogspot.com/

Introduction

I am pleased to present yet another practical title on breeding and raising livestock.

The work is in the Public Domain and is re-printed here in accordance with Federal Laws.

As with all reprinted books of this age that are intended to perfectly reproduce the original edition, considerable pains and effort had to be undertaken to correct fading and sometimes outright damage to existing proofs of this title. At times, this task is quite monumental, requiring an almost total "rebuilding" of some pages from digital proofs of multiple copies. Despite this, imperfections still sometimes exist in the final proof and may detract from the visual appearance of the text.

I hope you enjoy reading this book as much as I enjoyed making it available to readers again.

Jackson Chambers

GROUP OF YEARLING CHEVIOT RAMS.
On Plenderleith Farm, near Jedburgh, Scotland.

THE CHEVIOT SHEEP.

BY C. S. PLUMB.

About 300 miles north of London, forming the dividing line between England and Scotland, lies a group of mountains and hills. These are not rough, ragged, stone-capped mountains, such as are familiar to the New Englander, but rather smooth faced instead, covered with grass and vegetation to their summits. These are the Cheviot Hills. Among them a few rise to some height, of which the Cheviot at 2676 feet and Carter Fell rising 1815 feet, are the most prominent points. These hills mainly prevail in the north part of Northumberland County, England, and in Roxburg County, Scotland. Writing of this region in 1796, John Naismyth says[*] the whole Cheviot region is naked and open, and is now an unbroken continuation of sheep pastures, except such cultivation as is made for the accommodation of the flock. He tells of "beautiful, smooth, low, verdant hills," "clusters of fine, smooth knolls, covered with sweetest verdure," and of "a great range of good pasture of a mixed nature." Yet, he also writes of less fertile parts, "which having lain long neglected, the surface water has preyed on the soil, destroyed the sweet verdure, and brought a growth of mosses in its place," and further, that "upon the southwest of Cheviot, the plain top of the ridge is covered with a coat of peat earth, in some places very coarse and miry," producing various kinds of moorish herbage.

It is in this region that we find the native home of the Cheviot sheep.

When we consider we have but very little knowledge of breeds of live stock prior to the time of the Revolutionary war, and that in fact but few breeds were chronicled in histories 100 years ago, we can realize that ancient literature will give us but little light concerning the origin of this breed. George Cully in 1789 published the first distinctive work on live stock husbandry, in which he says,[†] "the original distinct breeds that I have seen, may I appre-

*Annals of Agr., Vol. 27, 1796, p. 177

†Observation on Live Stock, Dublin, 1789, p. 168.

hend, be reduced to seven." And in Cully I find my earliest reference to the Cheviot.

The tale is told by some that the Cheviot is descended from sheep that escaped from the vessels of the Spanish Armada, when they were wrecked upon the rocky shores of Britain over two centuries ago. This tale, however, has also been given in explanation of the origin of yet other breeds, so that coming from the sea, we may accept it as somewhat fishy in flavor and quality.

From all we can learn, this breed of sheep has been grown in the Cheviot Hills from very early times.

Naismyth, writing of the Cheviot hills in 1796, says* the sheep fed here were formerly called the long hill sheep of the east border, but are now better known by the name of Cheviot sheep.

Culley, seven years before Naismyth, describes the "long sheep," as he terms them, as having long, thin carcasses, thick planted, fine tender wool, and with white faces and without horns. Some few of these, he says, are speckled in the face and legs. They are "a kind of sheep, in my own humble opinion, very ill calculated for a mountainous country."

In 1784, Arthur Young, who perhaps is the most famous agricultural writer and author England has ever produced began the publication of his well known Annals of Agriculture. In the earliest volumes of this work, I find no reference to the Cheviot.

On November 10, 1791, David Irving addressed a letter to Sir John Sinclair, Chairman of the Society for the Improvement of British Wool, in the Annals of Agriculture, on the subject of "experiments with the Cheviot breed of sheep." In this he says, "having tried many experiments with the Cheviot, or as we call them in this country, *Long* sheep, and being convinced of their superiority over the Linton or Short breed for farms in the hilly part of Scotland, I take the liberty of sending to you and to the Society, an account of those experiments, in hopes that it will tend to remove the prejudices of such store masters as have not had an opportunity of trying the excellence of that breed which you have so properly recommended." He then goes on and details some of this work. In 1777 he bought forty Cheviot rams of the best sort on the farms of the Duke of Buccleugh and Sir James Johnstone, and

*Annals of Agriculture, p. 182.

crossed them with 1612 ewes of the Short or Black Faced breed. He states that these sheep were reckoned dear, and when his neighbors and kindred "saw me trying this new breed, they were very hard upon me, 'for pretending to go out of the good old way; for changing the good hardy for the soft long sheep,' and so forth." So he was frightened and gave up the experiment for eight years. But in 1785 he began the work in earnest, buying 134 Cheviot ewe lambs. In 1786 he crossed 420 black faced ewes with Cheviot rams. This work gave satisfactory results, and he secured one shilling higher for Cheviot blood than black face from the same pasture, while the wool which could have brought only 22£ 15s. from the old stock, sold this year at 48£ 19s. In 1787 he took the farm of Polmoodie, one of the highest lying farms in Scotland. He bought 1410 black faced sheep, 1079 of which he crossed with Cheviot rams. After giving the figures of his clips for 1788, '89, '90, he says: "Thus, I have brought the value of wool produced on this farm from 51£ 10s. to 115£ 7s. He notes that while his wool brought from 9 to 10s. per stone, farmers in the Cheviot region received 18 to 20s. per stone, and even a guinea, on account of having perfected their experiments. He also says, "The carcass I have brought to such perfection, that, as markets go, it will not be easy to add above a few pence to its value, at the same time with equal weight, the butcher can always afford to give from 9d to 1s. more for the Cheviot breed, on account of the better quality of the skin." He tells that farmers in the Highlands are prejudiced against the breed, and believes removing this prejudice will make the breed more valuable.

The same year David Irving reported on his experiments, the Society for the Improvement of British Wool submitted a plan* to the public for accomplishing the purposes of the Society.

In this the prevailing breeds of sheep are discussed. Here the statement is made that of all the breeds for the hilly parts of England or Scotland, the Cheviot, or South Border breed, is by far the most valuable. A special investigation was made of the breed in its native home by two of the Directors, Sir John Sinclair and Mr. Belches. They state that perhaps no part of the whole island where at first sight a fine wooled breed of sheep is less to be

---

*Annals of Agriculture, 1791, Vol. XVI, p. 428.

expected. Many parts of the sheep walks in those hills consist of nothing but peat bogs and deep morasses. During the winter their hills are covered with snow for two or three and sometimes even four months.

These sheep are long bodied. They have in general 14 ribs on a side. Their shape is excellent, and their fore-quarter in particular is of a full and proper weight. Their limbs are of a length to fit them for traveling and to enable them to pass over bogs and snows, through which a short-legged animal could not well penetrate. They are white faced, and have rarely any black spots on any part of their body. They have a closer and shorter fleece than the black faced, which keeps them warmer in cold weather, and prevents either rain or snow from incommoding them. They are excellent snow breakers, and from their habit of scraping the snow of the ground with their feet, have obtained the name of "snow breakers." They are, it is said, less subject to diseases than the common black faced breed. They sell at a good price for feeding. The weight of draught or cast ewes, which fetch 16-20 s. each, and wethers 14-16 s., and when fed at four years old is from 17-20 pounds per quarter. Lambs for feeding sell for about 7 s. apiece. From eight to nine fleeces make a stone of 24 pounds weight. The Directors cannot hesitate to recommend a trial of this breed to all the sheep farmers in the hilly parts of England and Scotland. For that purpose they have already purchased 50 rams and 100 ewes, which they propose to deliver at 36 s. the ram and 20 s. the ewe, in every district where application is made for that purpose by any active and intelligent improver. They state that there are now 30,000 to 50,000 of this sort brought to very great perfection. Of these Mr. Scott's at Lethem, Mr. Laing's at Plenderleith and Mr. Marshall's at Blindburn and Mr. Redhead's at Chatto, all in the neighborhood of Jedburgh, are among the hardiest and best. They state that the progress made in improving this breed during the past 20 years, especially relating to wool, is in the highest degree satisfactory. About 20 years ago the stone of wool required 10 fleeces and the wool sold at only 8 s. per stone. Eight fleeces now weigh a stone and the price is more than double. This committee believe that this breed may be improved by the use of foreign blood, and recommend crossing with Spanish sheep. They say the wool of this breed wants, 1st, to be finer in the pile; 2nd, shorter in staple, so as to make it fitter for clothing; thicker in the coat, so as to keep the animal warmer, and lastly more equal in point of quality, so that the whole fleece may be as nearly as possible the same. These are qualities the Spanish breed possesses superior to any other, and if the hardiness, the excellent carcass and the other advantages of the Cheviot breed are united to these properties or the Spanish, "*hill* sheep are brought to their greatest height of perfection." The cross between the two breeds has already been tried, and in so far as it is possible to judge from the appearance of the lambs this season, the experiment has answered completely.

In 1793, writing on Midlothian Agriculture in the Annals, George Robertson says, in speaking of the sheep there: "In the lower parts of the country, a better kind, on some farms has, however, been introduced, principally of the Cheviot breed, in some cases crossed with the Hereford, in others with the Bakewell species" (Leicester).

John Naismyth's sketch of the "The Cheviot Hills" in the Annals* in 1796 contains the most comprehensive and perhaps authoritative reference to Cheviot sheep that was written of the breed in its native home. Referring to this breed he says they are well polled and smooth faced, their fleeces unmixed white, and legs and faces either white, or somewhat mixed with black or brown, and this mixture on the face is always of a darker shade towards the nose. Those with the black mixture on the face are said, by some, to bear the finest wool. Their heads and ears are finely shaped; their countenances mild and pleasant. The whole figure is generally regular and well proportioned, but there are individuals in some flocks with rather thin shoulders, with legs too long for the size. The body, but more especially the tail, is longer than that of the Black Faced sheep, and hence probably has arisen the distinction of short and long sheep. The fleece is generally close, even and full topped; and the wool soft and fine, of from two to three and one-half inches in length. The same kind of polled sheep have fed in this district for time immemorial; nor does anybody allege that they were even natives of any other region.

*Annals of Agriculture, Vol. XXVII, 1796.

Formerly they are said to have been lank and gibletty; the back lax, the legs and neck long and slender and the shoulders thin. The fleece, though fine, was open topped and the breech hairy, and the lambs weak and thinly covered. About thirty years since, says Mr. Naisymth, which would be about 1754, Mr. Robson, a farmer of great professional knowledge and attention, is said to have been the first to attempt to remedy these defects. He brought rams from the wolds of Lincolnshire, to copulate with his ewes, by which the carcass and figure was much reformed, and by repeated crossings, obtained a highly improved stock. Naismyth says that the general practice has been to select a ram from any neighboring flock of the most approved shape, and soft, close and equal fleece, which is put to copulate with a few select ewes of the flock. From the produce of this connection, the most perfect lambs are picked to be used as breeding rams, when they are a year and a half old. This is repeated from time to time, and it is a rule not to use the same rams more than two years.

Sir John Sinclair, an eminent early agricultural author and man of prominence, who has already been referred to, took an active interest in this breed and stocked his own farm at Caithness, Scotland, with it. Writing in the Annals in 1793, he says:

Of the different breeds that are kept on the mountainous districts of this island, the Cheviot seem to be the best calculated of any perhaps hitherto known for such a pasture (mountainous); uniting hardiness, quality of wool and excellence of shape and mutton, and possessing that length of limb and body, which enables them to travel without difficulty, either in quest of food, or to a distant market. He states that the great object in regard to the Cheviot breed is to diminish the quantity of coarse and to increase the quantity of fine wool in the fleece as much as possible.

At this time, Young recommended what to-day would be considered a unique way to make the merits of a breed known. This was to appoint public agents at places where the breed is found in greatest perfection, for the purpose of encouraging the farmers in the neighborhood to attend more to the improvement of their flocks, and of corresponding with all those who might be desirous of purchasing at as little expense as possible. It was only by the means of such a plan that the knowledge of the Cheviot breed became so rapidly extended all over Scotland. A flock of 50 rams and 100 ewes were sent by the British Wool Society from the borders of England and Scotland to Caithness, about 350 miles, without the loss of a single sheep. They were sold to from 40 to 50 different people. The flock kept one route and every person got the ones intended for him. The cost of this, for expense, of driving 350 miles, tolls, ferriage, maintainance, shepherd, etc., was only about 1 s. 1¾ pence (27cts.) each.

The influence of the British Wool Society materially assisted in popularizing the Cheviot, and the breed gradually became distributed over Scotland and the highlands of England. Even before 1793, the Cheviot was introduced into the county of Sutherland in northern Scotland, and Mr. Kerr Richardson in the Annals for 1793, states that "the experiments which have been tried in that county, with the Cheviot breed, have thriven beyond the most sanguine expectations of those who have made the trial."

Prof. Wilson, an eminent Scotch agricultural authority of the early part of this century, states* that Cheviot mutton is excellent, and the carcass weighs from 12 to 18 pounds per quarter. The weight of the fleece is about three pounds. He calls it a hardy mountain breed, thriving well on waste and sterile lands. The Cheviots, he says, are characteristic of the hilly districts of the northwest parts of Northumberland, and are also bred on the hills around Cheviot, from which they derive their name. Their wool is in great demand and brings a high price. It is important to note here that he states some 50 years after it has become well known as a breed, that it has been much improved of late years, though there is still a want of depth in the fore-quarters, and of breadth, both there and at the chine.

David Low, Professor of Agriculture in the University of Edinburgh, in 1841-42 published a monumental work on "The Breeds of the Domestic Animals of the British Islands," illustrated by many large colored plates of great artistic merit of the different breeds. Plate eight in volume two is of a Cheviot ewe, bred by Mr. Thompson of Attonburn, County of Roxburgh. This represents an animal pure white in color, with smooth head, fairly large ears, prominent eye and nose and with black nostril. The body has a

*Quarterly Journal of Agriculture, Vol. II, No. 9, 1-29-31.

trifle more of length than we would desire to-day, and there is hardly the spirited carriage that is characteristic of the Cheviot as I have seen it both in England and America.

In writing of this breed in the work referred to, Low says that the body is very closely covered with wool, which is short and sufficiently fine for the making of certain cloths. Two shear wethers, when fat, may weigh on a medium from 16 to 18 pounds the quarter. The ewes are usually reckoned to weigh from 12 to 14 pounds the quarter. The mutton is very good, though inferior in delicacy to that of the South down and Welch sheep, and in flavor to that of the Black Faced Heath breed. The natural form of the sheep is, like that of all mountain breeds, with a light fore-quarter, but this character is removed by the effects of breeding, and the modern Cheviots are of good form. They are larger in the lower countries, where a supply of turnips can be given; they are lighter in the more elevated tracts, where artificial food is scanty or wanting. The breeders adopt the kind of animal which is suited to the pastures, preferring a shorter legged, larger sheep for the lower farms, and one of lighter and more agile form for the more upland and colder. The Cheviot sheep are of quiet habits, possessing, indeed, the independence of a mountain race, but having none of the indocility which distinguishes some other races. They are exceedingly hardy, their close covering of fine wool enabling them to resist the extremes of cold.

Low states that the Cheviots have spread over a large extent of country. In the southern Highlands in the forties, they had large-ly supplanted the Heath breed, and covered the elevated moors, formerly occupied by the Black Face Highland sheep. They had been carried to the extreme north of Scotland and to the west of England and Wales. He states that the extension that has already taken place of this hardy breed must be regarded as having been of singular benefit to breeders and the country. In its native country of the Cheviot hills, it has been cultivated with great care by a class of breeders inferior to none in the kingdom for intelligence and enterprise. Low notes that the wool of this breed weighs about 3½ pounds the fleece. It formerly used to be employed for the making of cloths, but from the extensive employment of the merino wool of Saxony and Spain, it is now scarcely employed for this purpose, and is prepared by the process of combing in place of carding for the coarser manufactures. The attention of breeders, too, having been mainly directed to the fattening properties of the animal, the wool has diminished in fineness, although it has increased in length and weight.

Undoubted the Cheviot went through a gradual change, as is shown through various writings between 1790 and 1840. Low brings this out in his reference to breeders increasing the fattening qualities, while the wool coarsened. Certainly 50 years before, a fine fleece was the first desire of the breeders, as is illustrated by using Spanish blood in flocks, as recommended by the British Wool Society. This change is further brought out by T. Rowlandson* in a prize essay "On the Breeds of Sheep Best Adapted to Different Localities," when he writes: How the Cheviot came to be classed amongst short wooled sheep by Sir John Sinclair and others, is to me inexplicable, except from the circumstance that people were content to use a coarser cloth formerly than at the present day. That the staple of the wool has been lengthened, and the wool otherwise become coarser, I am prepared to admit, but certainly not to the extent that the fleece has been changed from a short felting wool to a combing quality. Sinclair, we remember, wrote of the Cheviot as long and leggy, but over a half century later Rowlandson describes it as "a handsome, compact sheep, not quite so 'leggy' as the Cotswold and Yorkshire sheep, notwithstanding which they are an active race, are famous foragers, and withstand the vicissitudes of the weather exceedingly well, more so than any of the breeds previously noticed. The vigor of this breed seems attested by all. John Wilson, Professor of Agriculture in Edinburgh University, in 1855, says† they are exceedingly hardy, and, although possessing all the vigor and constitution of a mountain breed, exhibit none of their restless habits and submit with great docility to the restraint of the lowland farms. Wilson states that the natural pasture of the Cheviot range has aided in the development of a larger framed animal than the other mountain districts, that the fleece averages about five pounds, and the wool is of medium length and quality, and that the Cheviots have been

*Jour. Roy. Agr. Soc. of England, 1849, Vol. X, p. 421.

†Jour. Roy. Agr. Soc. of England, 1855, p 231.

crossed successfully with the Leicester and Southdown, and in both cases the produce has been satisfactory, showing an improvement in the carcass, the weight and quantity of wool and an aptitude to fatten at an earlier age than the pure breed.

Henry H. Dixon, in 1866, in a prize essay on the "Mountain Breeds of Sheep,"* says it is to the Robsons of Belford, who were flourishing when the century began, that the earliest improvement of Cheviots is generally allowed to be due. Their rams were all bred on the Cheviot ranges, and 70 or 80 of them would sell and let for about 700£ ($3,500), when they were marshalled each year in the great barn. It was said that there was a cross from Dishley in the flock. He tells of Mr. Reed who left the south side of the Cheviots to go to Sutherland, where he had as many as 18,000 Cheviots upon a farm 18 miles by eight, and that he turned over 2,000 three year old wethers and 1,500 cast ewes one September to a great Hawick salesman.

The southern Cheviots, according to Dixon, are brought up more artificially than those of the north, and so it is a question whether they are as hardy and active as the northern bred. Still most of the prize takers are of the south. Dixon notes that the most improved type of Cheviots, like Mr. Brydon's (for whose rams between 100 and 200 guineas were given at one of his biennial Beattock sales), have good Roman nosed heads, flat crowns covered with hard white hair, and "that cock of the lug and glint of the eye," which tell of mettle that will make them hunt the hill for food, and not hang listlessly round the hay hecks after a storm. They have also a fine "park ranging neck," rather Leicester-like girth and a width between the forelegs, light and clefty bone and plenty of wool under the belly, as well as on the arms and thighs. A good forearm or "butchers grip" is as great a point as white legs and a black nose; and the horned rams are thought more hardy, though they are often coarser in the coat.

It is a rare thing to cross breed Cheviots and Black Faces, says Dixon, but when this is done the Cheviot ram should be used, for if the cross is taken otherwise, the lambs are inferior both in shape and bone. The third cross of the Cheviot generally obliterates

*Jour. Roy. Ag. Soc., 1866, p. 360.

every trace of Black Face, except perhaps, in the grey shade of the legs and the kemp.

The Cheviot underwent great popularity, until the winter of 1859-60, which was of terrible severity in England and Scotland. At this time the Black Face demonstrated a greater hardihood than the Cheviot so that on the more exposed Highlands, the Cheviot is nearly altogether displaced. This I found to be the case in my visit in Scotland in 1897. In the upper Highlands, about the only breed to be seen, was the Black Face, and I saw them eating the heather in isolated numbers nearly on the summit of Ben Lomond, over 3,000 feet high. Referring to this hardiness of the Cheviot, Mr. Murray writes:* In the rainy climate of the far north the short and fine staple of the Cheviot wool is not so well suited to defend the skin of the animal from wet, as the long shaggy fleece of the Black Face, besides the Cheviots are more liable to be attacked by rot than the Black Faces, even when both are grazing on the same grazing, in consequence of their preferring the low boggy parts for the sake of shelter, while the Black Faces invariably prefer the dry, bare hights for their beds: yet even here the use of a Cheviot ram with Black Faced ewes has been successful, giving an extra value of five shillings to the lamb. In all the border counties, on the medium high ranges, where the climate is not too bleak and severe and some portions of turnips can be got in winter, the Leicester-Cheviot cross has answered admirably.

The inability to cope with the Black Face, in passing successfully through the severest winter weather, turned the scale in favor of the Black Face in regions where heretofore the Cheviot had more than held his own. Wrightson, in his work on Sheep,† says that on the lower and grassy slopes of the mountains, the Cheviot sheep maintains his position; but on the higher and less accessible tracts, where heather takes the place of grass, the Black Faced breed is best.

The type of Cheviots is rather different to-day from what it was 100 years ago, as has already been shown, and as we might expect. The form, style and character of wool have changed, and certainly improved. Professor Robert Wallace in 1893, among other things, gives these as features of the modern Cheviot of England and

---

*Jour. Roy. Ag. Soc., 1867, p. 570.
†Sheep: Breeds and Management, 1895.

Scotland.* The face and legs should be well covered with short, hard, wiry, pure white hair, which should extend over the ears and well back over the head. The horns in the rams, though not always present, are not objected to, being considered a sign of hardiness, if "clean," not thick and ringed or rough like those of the Black Faced breed. The nose of the ram is somewhat arched or Roman. The nostrils are black, and the eyes dark and very full and bright. The wool is moderately long, and should be close set, and neither open nor curly, but straight and free from "kemp" or dead hairs, covering all parts of the body well, including the belly, breast and legs down to the knees and hocks. A good average clip for ewes is 4½—5 pounds of washed wool. The tail is long, and should be very rough. It is cut so the point reaches the hocks. The shoulders are high and sharp at the withers. The fashionable form of Cheviot is now shorter, smaller, and more compact and the wool closer and thicker set than formerly, since a series of bad seasons, down to 1879, showed that the larger varieties with loose open fleeces, were not so hardy. The old, original Cheviot was a very close coated, short wooled and remarkably hardy sheep—even more hardy, it is recorded, than the Scotch Black Faced breed, a state of things which is now far from being the case. The ewes are good milkers. Cast ewes, fed on turnips for from 12 to 14 weeks, and wethers from the hills, at three years old, weigh, killed and dressed, 60.70 pounds; wethers a year younger, and finished on turnips, get up to about the same weight.

In early times, when Cheviots were quite fashionable, good prices prevailed for them. On September 13, 1865, 169 Cheviot rams sold at public auction at Beattock, by Mr. Oliver of Hawick, realized 2484£ 10s. ($12,422.50), which is the best proof we can give of the high estimation in which the breed was held by the flock masters of the border counties. One five year old ram sold for $190.00, one four year old for $570.00, one three year old for $775.00 and one two year old for $605.00.

A most interesting illustrated article, "A Day in the Cheviots," in the Live Stock Report in September and October, 1898, by Mr. A. S. Grant, Associate Editor of the North British Agriculturist, gives one an idea of the Cheviot of to-day in its native home, and

*'Farm Live Stock of Great Britain,' 1893. p. 217.

of the Cheviot country and shepherds. Sheep, he says, are everywhere in the great pastoral districts of the south-east coast. Up the steep hillsides they are as thick as they are on even the rich haugh lands by the railway side. Mr Grant's day was spent at Plenderleith, which lies at the very head of Oxnam Water, in the center of a famous group of Cheviot farms. On one side is the widely known farm of Under Hindhope, so long tenanted by the late Mr. Thomas Elliot, the famous breeder and improver of Cheviot sheep, and still occupied by his son, Mr. John Elliot, whose skill and ability in the breeding and bringing out of Cheviots is certainly not inferior to that displayed by his renowned father. Few breeders in fact, have been so successful in the show yard as Mr. John Elliot, and at the present moment his flock takes rank as one of the best flocks of the breed in existance. Nearly all the flocks in the country are more or less indebted to Hindhope blood, one of the most successful rams ever used at Plenderleith, Awfu' Sandy, having been bred there.

Mr. Grant says the Cheviot is *par excellence* the sheep of the Border districts. The original Cheviot, was, however, a very different animal from the thick, compact, active, finely wooled specimen of the present day. He was then comparatively small and scraggy in his frame, light in his bone, and with brownish colored head and legs. Now the head and legs are completely white, the frame thick and wide, and the gigots well filled. The well bred Cheviot of to-day is in fact perhaps one of the finest examples of combined energy and grace to be found in the whole ovine economy. He carries his head much higher than the border Leicester, which is also a very graceful sheep, and has at least twice as much fire in his eye. Few sheep can beat him in scaling a hillside, and none certainly can assume a more defiant or striking attitude. The Cheviot has also been very much improved in his mutton, his light square carcass cutting up to almost no waste, while his flesh almost rivals the mutton of the Black Face in flavor.

Speaking of the management of a Cheviot farm, Mr. Grant says that it is on the whole exceedingly simple. Generally speaking the sheep go at large over the farm during the whole season, individual animals rarely taking a wide range. The area required for each head of sheep varies from two to four acres. In some cases

where extra feeding is to be given, the rams are kept separate from
the ewes, but generally they are allowed shortly after weaning to
graze together again.   Ewes have their first lambs in April at two
years old and are sold as casts at five and six years old, being re-
placed by the best of the ewe lambs.   Cheviot casts are invariably
sold for producing a crop of lambs by Leicester rams.   As a rule,
the cast ewes, the wedder lambs, the small ewe lambs and the wool
constitute the whole produce of the farm.   This applies to the Cheviot farms in the south of Scotland and the north of England.

A history of this breed of sheep is really its English and
Scotch history.   The breed has only been known in America in a
very limited way, and but few sheep breeders outside of New York
have ever seen specimens of Cheviots in America or would know
them on sight.

So far as I have been able to learn, Cheviots were first brought
to the United States in 1838, and by Robert Youngs of Delhi, Dela-
ware county, New York, though they were imported into Canada at
an earlier date by Mr. Pope of Cookshire, Quebec.   In 1842,
George Lough of Hartwick and a Mr. Davison also imported
some to the same county.*   These became widely extended over
the southern central counties of New York, and especially Otsego,
and are mainly the foundations of stock in the American Cheviot
flock book.   From that time on, this has been the most prominent
Cheviot breeding ground in America.   Later importations were
made to this region by George Lough and Mr. Youngs.   E. J.
Bruce of Ketchum, N. Y. and William Curry of Hartwick,
imported "Willie the Wist," No. 1 imp. and three ewes in
1887, and William Ralph of Markham, Ontario, imported at the
same time.   In 1890 William Ainslie and Thomas Ainslie
and Son, of Hartwick, N. Y., imported from John Elliot of
Hindhope, Scotland, "Hindhope" No. 5 imp. and "Prince Davie,"
No. 6 imp.   Later on George Lough Jr. & Son of Hartwick, imported "Lady Robson" No. 16, imp. from John Robson of Newton,
England.   The only State fair in America, where Cheviots are a
prominent show, is that of New York.   In 1885, I think it was, in
the company of Col. F. D. Curtis, then a prominent agricultural
worker and stockman of New York, now deceased, I visited the
sheep pens, and there I made my first acquaintance with the beauti-

Sheep Industry in the United States, 1892, pp. 371, 396 and 414 ful, stylish Cheviot. We spent a liberal share of our time in the Cheviot pens, and Col. Curtis was very emphatic in his approval and praise of this breed. These earlier New York breeders claimed that the Cheviots were quiet in disposition and easily fenced and controlled. The rams sheared 8-12 pounds of wool, and the ewes four to eight pounds, well washed. The rams sold for $20 to $40 each. These flocks have been maintained with much purity for many years John Curry and William Ainslie settled in Hartwick, Otsego Co. and became Cheviot breeders along in the early fifties, and their descendants are now prominent breeders of Cheviots. Many small but excellent flocks exist in or about Otsego county to-day.

Cheviots were introduced into Washington County, Pennsylvania, by Thomas M. Patterson, of Patterson Mills, who in 1889 purchased a few heads in Otsego County, N. Y., since which time three small flocks have been brought into the county.‡ The conditions there seem favorable to their development. In 1891 Mr. Patterson's flock of 50 head averaged eight pounds of wool each, the wool being eight inches long. One ewe, three years old, weighing 196 pounds, clipped 10½ pounds of wool. Twenty ewes dropped 32 lambs.

In 1845, T. J. Carmichael imported three rams and six ewes to Wisconsin for his farm at Lake Mills, Jefferson county. These sheep were large and very fine, the fleeces quite as heavy and the wool nearly as long as the Leicester. The rams were bought of the flock of James Oliver, Bothwick Bray, and the ewes from Charles Scott of Roxburghshire, Scotland. This breed evidently did not become well-known in the State.

Cheviots were first taken into Illinois in 1888, when E. Pumphrey brought 10 ewes and one ram from the flock of E. J. Bruce, of Ketchum, N. Y. The next year they lambed 150 per cent.

Some time prior to 1838, a Mr. Pope, of Cookshire, Province of Quebec, imported Cheviots to Canada, and he later on made other importations. Since then a few flocks have been established in Canada.

*Sheep Industry in the United States, 1892.
‡bid . p. 520.

Cheviots were first brought to Indiana in 1891 by Howard H. Keim, of Ladoga, Montgomery county, who purchased 68 head of rams, ewes and lambs from the best flocks about Otsego county, N. Y. This is now the best known Cheviot flock in the west. Numerous other flocks have since become established in Indiana, so that now this State ranks only second to New York in the importance of its Cheviot flocks.

At the present time, there are also small flocks of Cheviots in Vermont, Michigan, Massachusetts, Iowa, Tennessee, Ohio and perhaps elsewhere.

There are two Cheviot Sheep Associations in America. On Jan. 28-29, 1891, the American Cheviot Sheep Breeders' Association was organized at Hartwick, Otsego Co., N. Y., the founders of the association being Ervin J. Bruce, Henry Van Dreser, William Curry and Howard H. Keim. This association has issued one Flock Book of about 160 pages. On March 23, 1894, the National Cheviot Sheep Society was organized at Indianapolis, Ind., with William Curry one of the Vice Presidents, and Howard H. Keim, Secretary. This society has published one flock book of 51 pages.

The Cheviot exists as one of the very best mutton and wool breeds of sheep known to-day, and no breed is better suited to the sheep uplands of America than the Cheviot. In a visit abroad in 1897, I talked with many sheep men, of different breeds, and everywhere the Cheviot was spoken of with the greatest respect and admiration. One of the best known Shropshire breeders of England spoke highly of the quality of Cheviot mutton, of its fleece, and of the general beauty of the breed.

There is no more beautiful breed of sheep than the Cheviot, and combining this with its other good qualities, in the hands of judicious, progressive breeders it should in future become one of the well known and established popular breeds of America.

The above is a lecture delivered before the National Cheviot Sheep Society by Prof. C. S. Plumb, of Purdue University, at its Annual Meeting at Indianapolis, Ind., January 4, 1899, and is published by order of the Society.   H. H. KEIM, *Secy.*

Brief Description of the Cheviot Sheep

BY THE SECRETARY.

A Cheviot ram, when arrived at maturity, weighs in good flesh at least two hundred pounds live weight. He has a lively carriage, bright eyes and plenty of action. His head is of medium length, broad between the eyes, well covered with short, fine white hair. His ears, nicely rounded and not too long, should rise erect from the head—low set or drooping ones are decided faults, but at the same time they should not be what are called "hare-lugged," that is too near each other, as that indicates a narrow face, which generally denotes a narrow body. His nose and nostrils must be black, full and wide open; his neck strong and not too long; his breast broad and open, with the legs set well apart. His ribs must be well sprung and carried well back toward the hook bones, as a long weak back is about the worst fault a Cheviot can have. His back must be broad and well covered with mutton; his hind quarters full, straight and square; the tail well hung and nicely fringed with wool. His legs must stand squarely from the body (if bent hocks, either out or in, and especially the latter, are looked upon as weakness); the bone must be broad and flat, and all must be covered with short, hard, white hair.

He will grow a fleece weighing 12 to 15 pounds of fairly fine wool, densely grown and of equal quality; coarseness on the tops of the hocks is a decided blemish. The wool should meet the hair at the ears and cheeks in a decided ruffle; bareness there or at the throat is inadmissible, and it should grow nicely down to the hocks and knees. The breast and belly are also well covered.

The same description, when modified, will apply to the ewe also, which will weigh one hundred and fifty pounds. Cheviots, when in a natural state, must grow finer wool, as hard feeding inclines to make it stronger, but it must be stiff and dense and not too short.

The perfect Cheviot is one which will live and thrive well on the hardest keep, and when taken to better ground prove itself equal to the occasion by growing larger and becoming easily fattened. The ewes are also great milkers, and very prolific.

RULES OF THE ASSOCIATION.

1. The name of this Society shall be "The National Cheviot Sheep Society," the objects of which are:

(a) The encouragement of the breeding of Cheviot Sheep and the maintenance of their purity.

(b) The establishment of a Flock Book of pure bred sheep, used in the past, and the annual registration of such lambs as are proven to the satisfaction of the Executive Committee to be eligible.

(c) The investigation of doubtful or suspected pedigrees.

(d) Any other business which, in the opinion of the Executive Committe, will be conducive to the best interests of the breed.

2. The officers shall consist of a President, First Vice President, V. P. for each State represented in the membership, and Secretary and Treasurer, each to be elected annually.

3. The business of the Society shall be managed by the Executive Committee, the members of which shall be chosen annually.

4. The President and Secretary shall be ex-officio members of the Executive Committee.

5. The membership fee of this Society shall be five dollars ($5.00), which sum shall entitle the member to the free registration of all animals bought or owned that are already registered in the Cheviot Sheep Society of Great Britain or The American Cheviot Sheep Breeders Association, provided said applications are made within six months of date of purchase. The registry fee for lambs up to and including twelve months shall be 40c. each, after which time 75c. shall be charged. Double fees to non-members. The transfer fee shall be 10c. The Secretary shall receive for his services 25 per cent. of the registry fees, and necessary stationery and postage. At the annual meeting the President shall appoint an auditing committee of two to audit the books and accounts of the Secretary and Treasurer.

6. A member, to be entitled to a vote, shall have paid all his dues, and must own and breed Cheviot Sheep.

7. These rules may be changed or amended at any regular meeting by a two-thirds vote. Members at a distance may vote by proxy under seal. The annual meeting shall be held in the month of January. Each member shall be notified of time and place of meeting by the Secretary.

For information or entry blanks please address the Secretary, enclosing stamp.

THE NATIONAL CHEVIOT SHEEP SOCIETY.

Organized March 23, 1894.

OFFICERS:

PRESIDENT,

PROF. C. S. PLUMB, ...   LaFayette, Ind.

1ST VICE PRESIDENT,

P. P. NOEL, ...   Rockville, Ind.

VICE PRESIDENTS FOR STATES,

T. N. CURRY, ...   Hartwick, N. Y.

U. S. MILLER, ...   Pulaski, Iowa.

H. C. DAVIDSON, ...   Elbridge, Tenn.

C. H. MARSHALL, ...   Vergennes, Vt.

SECRETARY,

HOWARD H. KEIM, ...   Ladoga, Ind.

TREASURER.

ISAAC LLOYD, ...   Russellville, Ind.

ARTIST,

L. A. WEBSTER, ...   Whiting, Vt.

EXECUTIVE COMMITTEE,

The President and Secretary ex-officio, P. P. Noel, Hon. D. W. Heagy, R. L. Ainslie, Isaac Lloyd, Wm. Curry, W. S. Crodian.

MEMBERS:

INDIANA.

1.   W. S. Crodian,   Fincastle.
2.   J. W. Brothers,   Estate, Morton.
3.   Wm. Hartman,   Fincastle.
4.   J. A. Guilliams,   Fincastle.
5.   P. P. Noel,   Rockville.
6.   Isaac Lloyd & Son,   Russellville.
7.   Jessie D. Ronk,   Ladoga.
8.   Howard H. Keim,   Ladoga.
9.   T. R. Lockridge,   Mace.
10. Prof. C. S. Plumb,   LaFayette.
11. Hon. D. W. Heagy,   Columbus.
12. J. Clayton Mahoney,   Ladoga.
13. S. M. Dunbar,   Bowers.
14. Luther Gardner,   Fincastle.
15. Grant Williams,   Fincastle.

NEW YORK.

16. Wm. Curry & Son,   Hartwick.
17. T. N. Curry,   Hartwick.
18. R. L. Ainslie,   Hartwick.
19. John Bowmaker,   Hartwick.
20. John R. Parr,   Hartwick.
21. A. H. Elliott,   Garrattsville.
22. John Lunn,   Edmeston.
23. Lee B. Webb,   Sugar Hill.
24. C. H. Ward,   Starkville.

IOWA.

25. U. S. Miller,   Pulaski.
26. Jeremiah Miller   Stiles.
27. J. C. Miller   Stiles.

VERMONT.

28. L. A. Webster,   Whiting.
29. C. H. Marshall,   Vergennes.

TENNESSEE.

30. H. C. Davidson,   Elbridge.